Mental Exercise for Dogs Book 2024

The Fun and Easy Guide to a Happier, Smarter, and Well-Behaved Dog + BONUS

[BONUS: Video Course + Audiobook]

Howard Briggs

Here is one of the biggest bonuses I promised you…

A full Audiobook on the **Mental Exercise for Dogs Book 2024**

To get access to it, kindly type in this link on your browser:

http://tinyurl.com/38d28jh7

OR

Scan the below QR code to gain access.

GET THE COMPLETE VIDEO COURSES AT THE END OF THIS BOOK

Table Of Content

Introduction

Imagine a future in which dogs are skilled at more than just retrieving sticks and racing after their tails; they are also involved in mental and sensory-challenging activities that improve their general wellbeing. Greetings from the amazing world of canine cerebral stimulation, where your pet may discover their inner genius and become the Albert Einstein of the dog park.

More and more dog owners and experts are realizing the many advantages of giving their four-legged friends mental exercise in recent years. Mental stimulation has revolutionized the field of canine care by improving cognitive performance and emotional well-being, decreasing anxiety and destructive behaviors, and increasing both.

However, what precisely is canine mental stimulation? Put simply, it's the process of stimulating your dog's intellect with different exercises that test their creativity, memory, and problem-solving abilities. It's similar to enrolling your dog at a canine college, where they may pick up new skills, work out riddles, and discover the world around them in a whole new manner.

What then is the significance of mental stimulation? To begin with, it may greatly enhance your dog's quality of life. A happy dog makes for a happy owner, and a cognitively active dog is happy. Giving your dog mental challenges helps to minimize boredom and harmful tendencies while also fostering a closer relationship between you and your dog.

Now that we know how important mental stimulation is for dogs, let's explore the several ways you may improve your dog's quality of life. The options are numerous, ranging from scent work and obedience training to interactive games and puzzle toys. We'll delve into the intriguing field of canine cognitive enrichment in this in-depth book, and you'll leave with the skills and information needed to discover your dog's inner brilliance.

So fasten your seatbelts and prepare for an exciting voyage into the realm of canine brain stimulation. Your dog will be grateful, and you'll be astounded by the amazing change that's about to happen.

The Importance of Mental Stimulation for Dogs

Similar to humans, dogs need cerebral engagement to maintain good general health. For dogs, mental activity is often neglected in favor of physical exercise. For their welfare, dogs of all breeds, ages, and activity levels need mental engagement, although to varying degrees. Even gentlemen need mental exercise! Let's examine the main justifications for why dogs need mental stimulation and—more importantly—how you may meet their demands.

Enhances virtuous conduct

Dogs may find alternative ways to amuse themselves if they're bored. Your dog may exhibit undesirable behaviors such as excessive barking, chewing shoes, destroying furniture, and/or uprooting the lovely flowers you just put in your yard if they don't get enough cerebral activity. It has also been shown that mental stimulation reduces canine hyperactivity and aggressiveness. Your dog may get more used to associating with humans and other canines by receiving mental stimulation. This might be particularly crucial if you have kids or other pets in your house.

Maintains dogs' wits

Dogs' mental and physical health are linked, just as in people. Your dog's mentality may be trained to avoid early aging. Since your fur kids don't live as long as we do—which is tragic—keeping them healthy is essential to creating more memories with them. Not only can mental stimulation help prevent dementia, anxiety, and depression symptoms in older dogs, but it also builds a solid basis for raising happy, intelligent puppies.

What you're capable of

Engaging your dog's brains via games, socialization, and obedience training may be very beneficial.

As your dogs get proficient at responding to orders, keep teaching them new ideas. Actually, obedience training helps dogs who are timid or afraid since it gives them confidence and makes them more accepting of other dogs and humans.

Do you know the dogs who get excited at the mention of "walk" or "treat"? Dogs can, in fact, link words to actions. Give your pets language lessons. Maybe play a game of hide-and-seek with them and teach them the name of their favorite toy. You can also make up an obstacle course, play fetch, positive light tug-of-war, and have them find a hidden goodie.

You may try turning on the TV or turning on some music for dogs that spend a lot of time alone. Your dog might benefit from sounds to keep them busy all day. Please remember that, depending on the type, size, and age of the dog, you should adapt the activities to suit its demands. A working breed, such as a Jack Russell Terrier, will need more exercise than a Basset Hound who is content to lounge about. Having said that, hiring seasoned caregivers to watch your dog while you are gone is a great substitute. Spend some time selecting a dog daycare where they can interact with both humans and other dogs. a location where kids may get proper attention and stimulation in a secure setting.

Additionally, keep in mind that engaging in mental stimulation exercises with your dog not only keeps them healthy but also helps to improve your relationship with them. It benefits both parties! You can encourage your dog's intellect to become active in a variety of ways. Use your imagination! While you're at it, give yourself a mental workout! Enjoyable!

Understanding Your Dog's Intelligence

Every trainer is aware that every breed varies significantly in intellect and trainability. But they also point out that dogs differ greatly from one another. Therefore, not all collies are winners, and not all winners are collies.

The person teaching the dog is mostly responsible for this. "If you are a good enough trainer, you can take a dumb breed and turn them into something really quite clever." The ease with which each breed may attain a certain level of performance and the highest possible standard that a dog of a particular breed may be expected to accomplish are the differences between the different breeds. Stanley Coren set out to determine which 133 dog breeds are the smartest. Since there was no data, he sent questionnaires to all judges in North America, and astonishingly, he received back about half of them. These are his research's findings.

The Brightest Dogs

Dogs in ranks 1 through 10 are the most intelligent in terms of working intelligence and obedience. In less than five (5) encounters, the majority of dogs will start to demonstrate a knowledge of basic instructions and will retain new behaviors without apparent practice. About 95 percent of the time, they comply with their handler's first order. Moreover, even in situations when the owner is far away, they react to orders in a matter of seconds.

1. Border Collie
2. Poodle
3. German Shepherd
4. Golden Retriever
5. Doberman Pinscher
6. Shetland Sheepdog
7. Labrador Retriever
8. Papillon

9. Rottweiler
10. Australian Cattle Dog

Excellent Working Dogs

Dogs in ranks 11–26 make great working canines. It should take five (5) to fifteen (15) repetitions to master basic instructions. Though they will become better with repetition, the dogs will recall orders rather well. Eighty-five percent or more of the time, they will comply with the initial order. There might sometimes be a little but discernible lag before the dog reacts to more complicated orders. Practice can help you reduce these delays. However, these breeds may be trained to perform effectively by almost any trainer, even if the handler lacks knowledge and patience.

11. Pembroke Welsh Corgi
12. Miniature Schnauzer
13. English Springer Spaniel
14. Belgian Tervuren
15. Schipperke
 = Belgian Sheepdog
16. Rough Collie = Keeshond
17. German Short-haired Pointer
18. Flat-coated Retriever
 = English Cocker Spaniel
 = Standard Schnauzer
19. Brittany Spaniel
20. American Cocker Spaniel
21. Weimaraner

22. Belgian Malinois
 =   Bernese Mountain Dog
23. Pomeranian
24. Irish Water Spaniel
25. Vizsla
26. Cardigan Welsh Corgi

Above Average Working Dogs

Working dogs in ranks 27–39 are superior than average. In fifteen (15) exposures, kids will start to demonstrate a basic comprehension of simple, new activities; nevertheless, it will take up to twenty-five (25) repetitions on average to achieve generally smooth execution. This group of dogs benefits from more practice, particularly in the early learning phases. When they pick up a habit, they usually stick with it. Their dependability depends on how much training they have gotten, but they will often answer to the initial order 70% of the time or better. Overall, these dogs behave much like the outstanding canines in the previous category. They just react a little less reliably, and there's often a noticeable delay between the instruction and the reaction. After a certain distance from their handlers, they won't answer consistently, and at very long distances, they could not respond at all. These breeds perform unquestionably worse when their handlers lack expertise or provide inconsistent or subpar training.

27. Chesapeake Bay Retriever
 =   Puli
 =   Yorkshire Terrier

28. Giant Schnauzer
 =   Portuguese Water Dog
29. Airedale
 =   Bouvier des Flandres
30. Border Terrier = Briard
31. Welsh Springer Spaniel
32. Manchester Terrier
33. Samoyed
34. Field Spaniel
 =   American Staffordshire Terrier
 =   Gordon Setter
 =   Bearded Collie
35. Cairn Terrier
 =   Kerry Blue Terrier
 =   Irish Setter
36. Norwegian Elkhound
37. Affenpinschers
 = Silky Terrier
 = Miniature Pinscher
 = English Setter
 = Pharaoh Hound
 = Clumber Spaniel
38. Norwich Terrier
39. Dalmatian

Average Working Dogs

Dogs in the 40–54 rank range are average when it comes to working and obedience training. After fifteen to twenty (15–20) repetitions, kids will start to demonstrate basic grasp of most activities throughout learning. Nonetheless, it will want twenty-five to forty (25–40) encounters to perform reasonably. These dogs will demonstrate strong memory with enough effort, however more practice during the first training phase would undoubtedly help. Without more practice, they can seem to lose the acquired behavior. More than 50% of the time, these dogs will react to commands on the first try, but real performance and dependability will rely on how much experience and repetition they get throughout training.

40. Soft-coated Wheaten Terrier
 = Bedlington Terrier
 = Smooth-haired Fox Terrier
41. Curly-coated Retriever
 = Irish Wolfhound
42. Kuvasz
 = Australian Shepherd
43. Saluki
 = Finnish Spitz
 = Pointer
44. Cavalier King Charles Spaniel
 = German Wire-haired Pointer
 = Black-and-tan Coonhound
 = American Water Spaniel
45. Siberian Husky
 = Bichon Frise
 = English Toy Spaniel
46. Tibetan Spaniel
 = English Foxhound
 = OtterHound
 = American Foxhound
 = Greyhound
 = Wire-haired Pointing Griffon
47. West Highland White Terrier

= Scottish Deerhound
48. Boxer
= Great Dane
49. Dachshund
= Staffordshire Bull Terrier
50. Malamute
51. Whippet
= Chinese Shar-Pei
= Wire-haired Fox Terrier
52. Rhodesian Ridgeback
53. Ibizan Hound
= Welsh terrier
= Irish Terrier
54. Boston Terrier = Akita

Fair Working Dogs

It is only appropriate to assess the working and obedience skills of ranks 55 through 69. When given a new command, it may sometimes take up to twenty-five (25) repetitions for them to show even a hint of comprehension, and it can take anywhere from forty (40) to eighty (80) instances for them to execute reliably. The habits could seem feeble even then. It can take them a long time and several tries to ultimately grasp the instructions and execute consistently and well. These breeds often behave as if they have forgotten what is expected of them if they do not get several additional practice sessions. To maintain performance at a respectable level, periodic refresher training sessions are usually required.

Only thirty percent of the time will these dogs react to the initial command at ordinary training levels. Even then, intimate training relationships are optimal for them. These dogs often give off the impression of being preoccupied, and they may only respond appropriately when they feel like it. These dogs' owners have to spend a lot of time yelling at them as they seem completely inattentive when their handlers are far away. Owners of these dogs often use the same justifications as cat owners to justify their animals' lack of response, asserting that their pets are "independent, aloof, easily bored," and other such platitudes. First-time owners should not own these breeds. Even an excellent dog trainer will struggle to get one of these dogs to behave with anything more than patchy dependability, but with plenty of time and tough but loving care, they may be trained to react properly.

55. Skye Terrier
56. Norfolk Terrier
 = Sealyham Terrier
57. Pug
58. French Bulldog
59. Brussel Griffon
 = Maltese Terrier
60. Italian Greyhound
61. Chinese Crested
62. Dandie Dinmont Terrier
 = Vendeen
 = Tibetan Terrier
 = Japanese Chin
 = Lakeland Terrier
63. Old English Sheepdog
64. Great Pyrenees
65. Scottish Terrier
 = St. Bernard
66. Bull Terrier
67. Chihuahua
68. Lhasa Apso
69. Bull Mastiff

The Most Difficult to Train

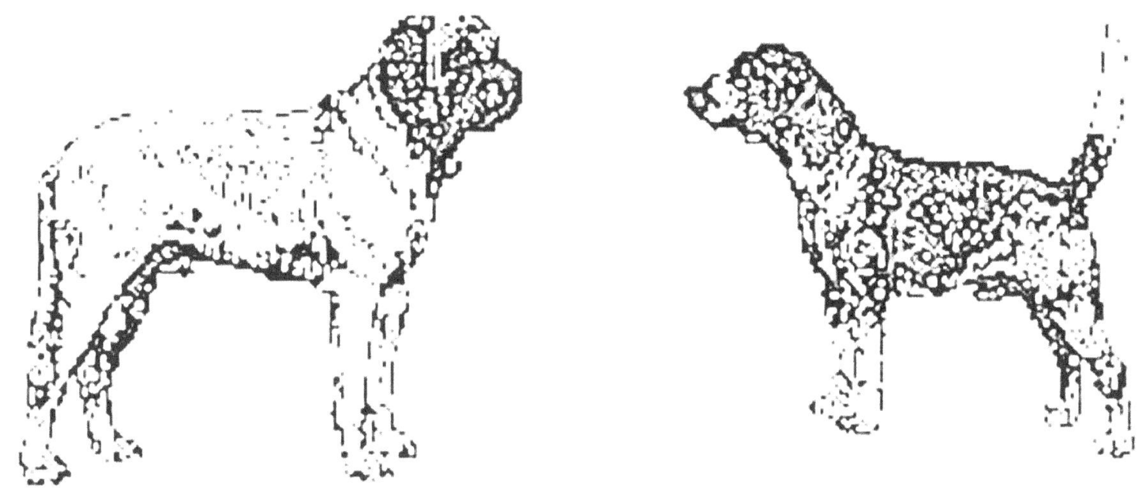

The breeds ranked 70–79 are thought to be the hardest to train, having the lowest levels of working and obedience intelligence. They can need more than thirty (30) or forty (40) repetitions during the first training phase before exhibiting the first hint of understanding what is required of them. Before any dependability is achieved, it is not uncommon for these dogs to need more than one hundred (100) repetitions of the fundamental practice exercises, sometimes spaced out across several training sessions. Even so, their performance could come out as shaky and sluggish.

Practice sessions need to be repeated many times after learning is attained, or else the dogs' behavior suggests that they never learnt the exercise in the first place and the training appears to vanish. Some judges stated that certain breeds were nearly impossible to train, while others hypothesized that the challenges most likely stemmed from the fact that the initial training and practice sessions were not prolonged enough with average handlers for the behaviors to solidify into ingrained habits. Even after forming a habit, many breeds continue to exhibit erratic responses. They sometimes turn away from their handlers, giving the impression that they are deliberately disobeying orders or defying their owner's authority. When they do react, they often do it very slowly and exhibit a lack of confidence or dissatisfaction with their actions. While they perform well under supervision, several of these dogs are unreliable when left alone. These need the most skilled and knowledgeable handlers of any breed.

70. Shih Tzu
71. Basset Hound
72. Mastiff
 = Beagle
73. Pekingese
74. Bloodhound
75. Borzoi
76. Chow chow
77. Bull dog
78. Basenji
79. Afghan Hound

However, what about crosses or mixed breeds?

Here, the dog judges, whose responsibility it was to evaluate the behavior of purebred dogs of purebred, were less certain. It seemed that judges, together with those who taught obedience lessons and served as trainers, could reasonably forecast and rate mixed-breed dogs. In general, they believed that a mixed breed dog would most likely behave like the breed it most closely resembles. A beagle-poodle hybrid will thus behave a lot like a beagle if it looks like one. Its behavior will be quite similar to that of a poodle if that is how it appears.

However, the majority of mixed breeds have certain traits and tendencies that are unique to both breeds, which has contributed to it. The likelihood that a dog's behavior will combine traits from both parents increases with the degree of physical similarity between the dog and its parents.

The Benefits of Mental Exercise

Mental and Physical Health Are Correlated

For dogs as for humans, regular exercise is the cornerstone of good health. Maintaining a healthy weight, enhancing cardiovascular health, and preserving muscle and joint function are all facilitated by physical exercise. The psychological well-being of a dog is greatly enhanced by these physiological advantages.

Psychostimulant

Dogs are bright, naturally inquisitive animals. By keeping kids physically active, you can stimulate their minds and keep them from becoming bored or developing harmful habits. A dog that receives mental stimulation is pleased and happy.

Social Communication

Dogs who are socially active tend to flourish. Frequent exercise gives dogs the chance to interact with humans and other pets, promoting healthy connections and lowering feelings of loneliness or isolation.

Chapter 1: Getting Started

With your animal companion, are you prepared to go off on an adventurous adventure? We'll go over the fundamentals of mental stimulation for dogs in this chapter and assist you in setting the stage for a happier, more involved dog friend. Now let's get started!

Assessing Your Dog's Current Mental Stimulation Level

It's important to determine your dog's current level of mental stimulation before you begin adding mental workouts to his life. Observe the following warning signs:

- Behavior: Does your dog engage in harmful activities, such as digging or excessive chewing? These might indicate a lack of mental stimulation and dullness.
- Does your dog have trouble focusing on duties or training exercises? This might be a sign that they need further mental exercise.
- Energy levels: Does your dog never stop playing or going for lengthy walks? This may indicate that they need more mental difficulties to exhaust themselves.
- Sleep patterns: Does your dog sleep longer or find it harder to fall asleep than usual? This may indicate that they need more mental stimulation in order to feel more at ease and satisfied.

It's time to establish some objectives for your dog's mental training regimen now that you have a clearer picture of their present state of mental stimulation.

Setting Goals for Your Dog's Mental Exercise Routine

For your dog's mental exercise regimen to be successful, you must set reasonable objectives. To help you get started, consider the following:

- Start modest: Start your dog out with easy activities and as he gains comfort and confidence, gradually raise the level of difficulty.
- Be dependable: Create a mental workout schedule and do your best to keep to it.
- Track development: To make sure your dog stays challenged and interested, monitor their development and change the objectives as necessary.
- Honor significant anniversaries: Remember to acknowledge and appreciate your dog's accomplishments, no matter how little! Maintaining their motivation and desire to study requires constant positive reinforcement.

After establishing your objectives, let's go to the following phase, which is selecting the ideal training setting.

Choosing the Right Environment for Training

The effectiveness of your dog's mental activities may be greatly influenced by the setting in which you do them. When selecting a training setting, keep the following things in mind:

- Choose a spot where there won't be many distractions, such other dogs, loud sounds, or strange people. This will assist your dog in maintaining attention on the current job.
- Comfort: Make sure your dog is at ease in the training area by providing a soft surface for them to sit or sleep on and a temperature that is appropriate for them.
- Size: Make sure the area is roomy enough for your dog to go around freely, without being too big to the point where they feel overwhelmed or sidetracked.
- Familiarity: If at all feasible, take your dog to a place they are used to, since this will make them feel more comfortable and secure throughout training.

You're well on your way to being an expert at mental stimulation now that you know how to determine your dog's current degree of mental stimulation, create objectives for their mental exercise regimen, and choose the ideal training setting!

We'll delve into the fascinating realm of fundamental mental exercises for dogs in the next chapter. Prepare to discover a wide range of captivating and intellectually interesting activities that will enable your dog to fulfill their greatest potential. Keep checking back!

Chapter 2: Basic Mental Exercises

Hide and Seek

Imagine this. Your dog is rushing around the house, full of energy, and unable to go for his regular walk since you're stranded inside on a rainy, freezing day. How can I burn off all of that extra energy? A traditional game of hide and seek. This is not only a great game for your dog, but it also exercises his mind and body as it requires him to find you, solve problems, and think. Playing hide-and-seek also helps your dog develop its sense of scent, which it will use to find you. We are not very good at smelling as humans. For dogs, however, scent is very crucial. Playing this game with the family is a terrific way to get everyone moving and entertained at the same time. The following instructions will help you educate your dog how to play hide-and-seek effectively.

First, let your dog see you hiding.

The advantage of this strategy is that it lets your dog track you down. Dogs search for their pack leaders nonstop. Make sure someone else is holding your dog's leash. While you go hide behind a piece of furniture, a tree, a shrub, etc., engage your dog in lively conversation. Tell the dog that you are hiding.

Step2: Give them something for locating you.

 Tell your dog to come find you. After removing your dog's leash, have your helper follow him. Your dog will probably rush to you because he can see where you hid and uses that information to determine where you are. This introduces the idea of hide and seek. Give your dog food or toys when he finds you as a way to say "thank you."

Step 3: Increase difficulty.

Up the difficulty a little in the game. More distance than before should be hidden. Make your hiding places more intricate, and keep your dog from seeing where you are. Allow the dog to explore and utilize his hearing, sight, and scent senses to help him.

Transfer objectives in step four.

Ask your helper to give your dog the "find" command. As you lead your dog to you, begin to replace yourself with someone else who hides. When you are the one leading your dog on his search, give him praise for locating the new individual.

Treasure Hunt

Remember that we are not discussing an actual treasure hunt activity. Even then, you never know what may occur if you teach your dog to search. Perhaps they will take you to a treasure that has been concealed for a long time.

However, you may enjoy yourself with your dog in the meantime while they grow in both their physical and mental abilities. The scavenger hunt is one of the games that may be played quickly.

Every dog owner knows that dogs are scavengers by nature and like looking for food. You may capitalize on their desire to help them increase both their physical and mental energies in order to get the treat. Once you know how to construct a treasure hunt, you won't need much equipment—even canine puzzles and homemade toys that hold scavenger hunt clues will work just well. Naturally, you should consult the veterinarian about safety concerns before beginning to play.

Take it gradually at first. Dogs are not all the same. Every dog is different when it comes to how they respond to the suggested treasure hunt ideas, regardless of how much they like eating. Take a pause if you see that your dog is becoming tense or hiding for whatever reason.

Use distance rather than forcing your dog to approach if they are terrified of anything, like a toy. Allow them to smell it and familiarize themselves with it at their own pace.

Treat your pooch with some goodies. For example, you may make a treasure trail with kibble or treats if your dog seems scared of the toy or doesn't appear interested in searching for it. Your little four-legged companion will eat the pieces as they go until they reach the prize they were supposed to locate. When toys smell great, even the most fearful dog will eventually get over their apprehension and get excited about the prey treasure hunt.

Remember that the goal here is to enjoy yourself and your dog, not to get them ready for an actual treasure hunt. The dog will be pleased if you are more at ease and tolerant with them. It is never going to work to force dogs to do anything they do not want to.

As the dog ages, you may introduce more challenging workouts. Some refer to this as a K9 hunt. You tell your dog to sit while you show them the goodie. Next, instruct the dog to search for the treasure by hiding the present someplace and making sure it is visible to the dog. Once your dog has mastered the first step, you may let them play pretend by pretending to store goodies in various locations while you are really giving them to them.

Puzzle Toys

Don't ignore your pet's mental health even if their physical health is the most important thing to take care of. New challenges and learning opportunities are very beneficial to dogs. Engaging in mental stimulation for your pet may help them shed extra energy, reduce tension, and increase mental endurance. Additionally, it may aid in the treatment of typical canine problems including separation anxiety and unruly behavior brought on by boredom. Just like people, dogs may improve their mental skills by playing with stimulating and interesting puzzle toys.

Benefits of Puzzle Toys for Dogs

Puzzle toys are designed to pique your curiosity and promote the growth of your problem-solving abilities. Including puzzle toys in your dog's daily routine can keep him occupied for hours, which will enhance his mental and physical health.

The following are a few of the main benefits of puzzle toys for dogs:

Reduce Boredom

There are times when you may not have the time or energy to play with your dog. Games and puzzle toys may relieve your pet's boredom and encourage autonomous play. Providing your dog with a way to release his extra energy may also assist prevent trouble stemming from boredom.

Stop Anxiety

Does your dog have separation or general anxiety? Puzzle toys could be useful for keeping your pet comfortable and peaceful. Dogs who stay occupied and concentrated on a task are less likely to get agitated in stressful circumstances, as when their owners leave them alone at home.

Promote Joyful Play

Puppies have a reputation for being troublemakers. They often dig, roll, lick, and chew anything or surfaces that catch their interest. Young pups may satisfy their curiosity with a safe and healthy exercise that comes with puzzle toys. When given access to cognitive toys, dogs are less likely to behave badly.

Prevent Dementia

Dogs that are elderly may also suffer from dementia, just like people. Typical signs of canine cognitive dysfunction (CCD) include irregular sleep patterns, involuntary barking, anxiety, house accidents, personality changes, or reclusive behavior. In order to prevent frequent problems like memory loss and diminished cognitive function, pets need to engage in mental exercise. Puzzle toys are a great way to help your dog think critically and practice problem-solving techniques.

Encourage Joy

Dogs may enjoy games and puzzle toys, which improves their general happiness and wellbeing. Not every dog likes standard dog toys like noisy toys or soft toys. For dogs of all ages, puzzle games provide a different kind of amusement.

Kinds of Dog Puzzle Toys

There are a wide variety of puzzle toys available for dogs on the market. Certain puzzles work better for certain dogs than others, and some of these games are harder than others.

One of the most popular puzzle toys for dogs has many concealed sections that hold treats or kibble. To uncover the food buried behind these puzzles, the dog must slide, topple over, or lift pieces. Owners of pets may increase the difficulty of this game by concealing only one reward inside the puzzle. The dog next has to locate the treat's hiding place using his excellent sense of smell.

Another popular puzzle item that makes dogs utilize their amazing sense of smell is a snuffle mat. On the play mat in the enrichment activity, food may be concealed among the strands that resemble grass. After that, your dog will have to explore the mat and utilize his innate hunting skills to locate the delicious morsels. Snuffle maps are a great way to exercise your mind and body.

Pets are also often advised to use licking mats because of their many health advantages. These mats work well for both cats and dogs and are great for dispersing natural, organic, or healthful foods. Pets may be soothed and calmed by licking mats as they explore all of the crevices and corners to find rewards. Additionally, by making your dog eat more slowly, licking mats promote healthy digestion.

Dogs may also like hide-and-seek toys as entertaining puzzle toys. These toys, which often come in plush form, include a bigger toy within smaller toys. A hide-and-seek toy may, for instance, have a big bee nest with little cuddly bees inside of it. By encouraging your dog to shake, move, and take the little toys out from within the big plush toy, hide-and-seek toys help prevent boredom.

A Few Points to Remember

Dog puzzles provide several advantages; nonetheless, there are some factors to take into account while using these interactive activities and toys.

- While your pet is playing with puzzle toys, keep an eye on them. Certain toys are not designed to be chewed on, and frequent biting or gnawing may cause loose parts that might choke a child.
- Seek for puzzle toys that are of a higher caliber, secure, long-lasting, and non-toxic, since they are more likely to survive severe use.
- Take care not to overfeed your pet while offering them puzzle toys that include food or treats. While interactive feeders are a terrific way to add excitement to mealtimes, be careful not to overfeed your dog.
- Puzzle toys should not take the place of spending quality time with your pet. Puzzle toys should never be used in place of affection and care; they should only be used to keep your pet busy for brief periods of time.

Learning New Tricks

A methodical approach to teaching your dog new skills.

Learning new skills from your dog may be enjoyable and beneficial for both of you. Dogs love to train because it gives them plenty of attention, mental stimulation, and rewards, of course. Bonding with your dog via training may be a fulfilling experience.

Any dog can pick up these five tricks with the correct methods and with practice.

Trick 1: Sit

"Sit" is one of the most often used and basic commands, and it's a wonderful first skill to teach your dog.

1. Grab a goodie firmly in your palm and use it to attract your dog's attention. Hold the reward in your palm just over your dog's head; it should be just out of reach for them, but not so high that they attempt to leap for it.
2. Give the order to "sit." Simultaneously, carefully slide your hand from your torso toward the rear and tail of your dog. As they pursue the reward, many dogs will naturally tilt their heads back and sit.
3. If your dog is sitting, respond with a "Yes!" as soon as their rear end reaches the ground. After giving your dog the reward, give them praise.
4. After a few tries with this technique, if your dog still won't sit, don't force them to sit by applying pressure to their behind. Rather, watch your dog as he normally behaves. As soon as you see your dog preparing to sit, praise them with a reward and say "Sit," "Yes," or "Good." They will gradually start to connect your vocal signals and positive reinforcement with sitting.

Trick 2: Shake Hands

Despite its spectacular appearance, this technique is really rather easy to teach. Adhere to these seven guidelines.

1. Hold a goodie in your palm to begin.
2. When your dog smells a treat, it will want to eat it. Hold your hand closed. The innate tendencies of most dogs will lead them to paw at objects that are out of their jaws.
3. Say "Yes!" and give your dog the reward as soon as they put their paw up to your hand.
4. Continue until your dog consistently and swiftly extends its paw to your closed hand.

5. Next, give your dog a flat, empty palm. Give them a reward when they place their paw on your palm.
6. Give your dog's paw more time in your palm before offering them a reward.
7. Next, provide a spoken cue—such as "Shake!"—right before you extend your flat hand.
8. Your dog will have mastered the trick after you go through these stages a few times.

Trick 3: Roll Over

Teaching your dog any skill requires a lot of repetition. This is particularly valid for the "roll over" ruse. Your dog will grow better the more times they finish these stages.

1. Your dog should begin by laying down.
2. Keep a reward close to your dog's nose and offer it to them. To have your dog raise their head and turn to the side in order to recover it, move your hand to the side, over their shoulder. Hand them the candy.
3. Give your dog another reward right away, this time without letting it go, and urge them to turn over and change their weight. Place the reward on the floor just out of your dog's reach so they have to roll over to receive it. Give your dog a reward and praise them if they succeed in doing this.
4. Give the order to "roll over" after a few successful rolls, then gradually reduce the treat.
5. Continue your regular practice. It could take some time for your dog to turn over with ease while eating only one goodie.

Trick 4: Play Dead

The "play dead" trick involves your dog rolling on their back, maintaining their posture without completely turning over, much like the "roll over" trick. Encourage your dog to "play dead" after they have mastered the "roll over" exercise.

1. Your dog should begin by laying down.

2. To get your dog to roll onto their back, give them a treat on one side of their snout and then move your hand with the reward to the other side of their body.
3. Tell your dog to "play dead" or "dead" while they are lying on their back. Sometimes it's helpful to add a physical gesture, like a raised palm, to go along with this spoken message. To hold your dog in place, place one hand on their chest.
4. Give your dog a reward and say "Yes!" after they have been resting on their back for a little period.

Trick 5: Take a Bow!

The ideal trick to do after teaching your dog all of his new abilities is a bow. Your dog will bend forward on their front elbows, contacting the ground with their chest, to make a bow. The back end of your dog will remain up in the air.

1. To begin, your dog should be upright with all four paws planted firmly on the ground. If they can rise at command, that's beneficial.
2. Treats should be held at the very tip of your dog's nose and moved gently down toward the ground, staying close to their body to entice them to come down until their back ends are still elevated and their front legs are on the ground.
3. Some dogs find this posture difficult at first. Try holding your dog's rear end in the air with one arm under their tummy to help them comprehend, and then use the goodie in your other hand to gently bring them down if they still don't understand. Continue doing this until your dog knows how to do the required movement.
4. Remain in the bow posture and use the reward to encourage your dog to come back up to a standing position. Once you respond "Yes!" or with a clicker, offer the reward.
5. Continue practicing and repeating. Add a cue word or phrase eventually, such as "Take a bow!" before you begin the act. Your dog will be able to bow on demand before you realize it.

Interactive Games

After utilizing its brain for ten minutes, a dog may be more exhausted than it is after an hour-long stroll. Letting the dog utilize its sense of smell is one of the best strategies. Making entertaining activities that keep your dog occupied and happy is really simple!

To ensure your dog doesn't overeat, it might be a good idea to utilize a portion of its food as rewards. This method of giving food will satiate your dog's innate need to smell things.

TURN MEALTIME INTO AN INTRIGUING SCAVENGER HUNT!

What you require: Sweets.

Disperse the goodies about the home so that the dog may scent them to locate them. To start simple may be a wise move. Allow the dog to see where the goodies are being placed. Lay them out flat on the ground without covering them up too much. You may eventually up the difficulty by:

putting the candies behind, under, and between things.
arranging the candies at different heights—on a chair, a shelf, etc.
extending the region under search to many rooms.
Keep the dog from seeing where the goodies are being placed.
You may do the same outside.

SNUFFLE MAT

You will need fleece blankets and a rubber matting.

Slice the fleece blankets into lengths of about 35 centimeters and widths of about five centimeters. Make a double knot to secure them to the rubber mat. You get to decide on the mat's size and the closeness of the knots. Once the rubber mat is finished, you may cover it completely with snacks.

Another variation that is simpler to construct is a braid. Cut three strips off of a kitchen towel or similar cloth. Create a braid and secure the ends with a rubber band or a knot. Stuff candies inside the cloth.

FROZEN DELICATES

Treats and water or moist food are what you need. A toy that can hold sweets, or an ice cube tray and freezer.

Since the food is frozen, the dog is taking longer to eat—ideally in a toy designed specifically for this purpose. Fill the desired item with either wet food or a combination of water and goodies. Once the mixture is frozen, place it in the freezer.

CANDY IN A BOTTLE

Treats, an empty bottle, and a drill or knife are required.

Your dog may enjoy an interesting game with an old plastic bottle!

One option is to put goodies inside the bottle, screw on the bottle top, then drill large holes in it at various locations. The idea is to get the dog to turn over and shift the bottle so that the goodies drop out. The dogs will find it simpler to remove the goodies from the bottle the larger the hole you create. Thus, it might be a smart idea to start with a larger hole and a simpler level of difficulty so that your dog can grasp the purpose of the toy. Dogs may abandon them if the task is too difficult! Then you have to simplify it.

Make sure there are no sharp corners where the dog might be hurt.

If your dog succeeds at this, you may attempt a more difficult variation. Fill a hole-free bottle with goodies. The goal is to get the dog to move the bottle so that the goodies fall out of the aperture, so don't screw on the cap. Start the process with your dog in mind by nearly completely filling the bottle with goodies. Once the dog is proficient at the job, you may reduce the amount of rewards they get.

A third variation involves perforating three bottles, inserting a stick through them, and suspending the shaft from a platform. Fill the bottles with goodies. The bottoms of the bottles are not pointing toward the earth. The object of this game is to get the dog to spin around and let the goodies fall out by hitting the bottles with its paw.

ENTERTAINING MUFFIN PAN

Treats, tennis balls, and a muffin tin are needed.

Place the candies in the pan's bottom. Cover them with tennis balls. The dog must maneuver the tennis balls in order to get the rewards.

You can play the same game using an egg carton if your dog is little.

Here's how our pals at McCann Dog Training go about things:

BLOCKED CARDBOARD CASES

Cardboard boxes, candy, tape, and finally newspapers are everything you'll need.

When it comes to versatility, the only limit is your creativity! Newspapers, for instance, may be crumpled and stacked in a small cardboard box. Make sure the staples are removed to prevent self-harm for your dog!

Treats should be dispersed throughout the box so the dog has to search for them.

For a somewhat more challenging variant, fill smaller food boxes with snacks and newspaper, tape them together, then place the smaller boxes inside a larger one. Tinier treat boxes may also be inserted into other items, such as a plastic bag holder. For an added challenge, you may also run paper towel tubes stuffed with candy through this one.

TOWEL PAPER TUBE

Paper towel tubes or toilet paper are required.

A toilet paper or paper towel tube packed with rewards may make great interactive games for your little dog.

After placing the snacks inside the tubes, fold the ends over. This may also be frozen or stuffed with newspaper if you want to make things harder.

Chapter 3: Advanced Mental Exercises

Agility Training

If you've ever seen dog agility on television, you know that it's a thrilling, fast-paced sport for both dogs and their owners. Even more enjoyable than it is to observe, however, is doing it with your dog!

Beyond physical advantages, agility fosters stronger relationships between owners and dogs and provides excellent cerebral stimulation, especially for hyperactive canines. For advice on how to get started at home if you're considering giving it a try with your dog, continue reading.

Canine agility: What is it?

Dog agility is an obstacle course that dogs and their handlers must finish in a timed, fast-paced manner. The dog must navigate a course's fifteen or more obstacles—which might include weave poles, tunnels, jumps, and ramps—in a precise order.

The advantages of training agility to your dog

The advantages of training your dog agility are many!

Excellent exercise: Your dog will become exhausted running, leaping, climbing, and weaving—all at a rapid speed. For dogs with a lot of activity and a desire for frequent exercise, agility is very beneficial.

Minimizes Boredom: Agility not only gives your dog a physical exercise but also cerebral stimulation that helps cut down on boredom and the negative behaviors that might accompany it.

Encourages positive behavior: Dogs depend on their owners to give them the orders they need to successfully finish the agility course. Teaching agility to your dog improves their attentiveness toward you and strengthens their adherence to your orders.

Strengthens the relationship between owner and dog: A lot of agility handlers talk about how team training has strengthened their relationship with their dogs.

Is agility a good fit for my dog?

Every adult dog in good health may do agility; however, extremely elderly or sick dogs should not perform agility. It's best to begin very gradually with a puppy and to finish a course only when the dog is between 12 and 18 months old. If your dog has arthritis or is prone to back pain, like certain Dachshunds or Basset Hounds, you may also need to modify or eliminate some of the features, such as leaps.

Agility is also beneficial for dogs with typical behavioral problems, such as the following, according to many pet owners:

Nervous or anxious dogs: Uncertainty scares anxious or nervous dogs, who often get stressed out by not knowing what will happen next. These dogs may benefit from agility classes as well as from routines and repetitive exercises. The sport rapidly becomes second nature to them, and they take pleasure in the repetitive patterns that reduce their tense or apprehensive behavior.

Working dogs with high energy: If your dog belongs to a working breed, you may find it difficult to provide them with adequate exercise to release all of their energy. Agility may be a physical and mental challenge for these extremely sensitive dog breeds, which can help exhaust them.

Dogs that struggle with concentration: Agility classes may be a good option for your dog if they are a wiggler or fidgeter who finds it difficult to concentrate, whether at home or in a training session. This will help them focus and redirect their energy.

Is agility something I should do?

It's crucial to think about if dog agility is a good fit for you and your dog since it should be enjoyable for both of you. Since agility may happen quickly, you should set up a regimen of strength, agility, and balance exercises for yourself if you don't already have a strong foundation of fitness. This will help you assist your dog throughout the training process.

Make careful to go slowly until you and your dog are both comfortable finishing a course.

Things to consider before beginning to teach your dog agility

Set a foundation of fundamental compliance.

Throughout the agility course, your dog will need to carefully follow your commands, so review the fundamentals of obedience with her. Use positive reinforcement training methods to teach her to sit, stay, and come. (Learning to stay is especially crucial with the teeter board, since your dog must remain still until the far end reaches the ground.)

Increase your dog's capacity for focus.

Ensure that your dog can focus on you in spite of other distractions. There could be a lot of distractions on the course, such as other dogs, loud sounds, and of course, all the exciting obstacles to play with! When you're out and about, practicing commands like "Watch me" or "Look at me" might assist with this.

Teach your dog to tolerate a variety of motions.

Before exposing your dog to the actual obstacles, make sure she is familiar with moving in unusual ways. Teach your dog to step on top of, climb over, and crawl through objects. She may also be trained to walk backwards and to position her paws in certain spots. By throwing goodies in the direction she wants to go, you may teach her how to spin an item around firmly and walk away from you or to the left or right.

At-home agility training for your dog

You could wish to construct your own agility obstacles at home to determine whether your dog loves agility before taking them to a class. Here are some pointers for building DIY obstacles and teaching your dog the fundamentals.

First, be sure to find out from your veterinarian whether jumping is appropriate for your dog. It is not advised for certain breeds to jump since they may be more prone to back issues than others, such Basset Hounds and Dachshunds.

Once your veterinarian gives the all-clear, you may make leaps by placing a piece of plywood to stop several books. Verify that if your dog knocks the jump over, they won't injure themselves. Step down at first, then gradually raise the leap height. Start the board one to two inches from the ground for larger breeds; for smaller breeds, you may want to start the board on the ground.

Tire Leap

For the tire leap, an old bike tire or hula hoop ring might be a useful tool. Just make sure your dog can fit through it without difficulty. First, let your dog go through the tire while you maintain a solid grip on it. It may be raised gradually until it is hanging from a tree limb.

Dog Trot

There are ramps at both ends of the elevated dog path. Consider using a picnic seat with wood pieces placed at each end to serve as ramps.

You may want to start with the end piece of the obstacle since some dogs can be reluctant to clamber up onto it. After picking up your dog, drop them down not far from the obstacle's conclusion. Dogs will often take the few steps necessary to get off. After that, you may increase the challenge until your dog is content with it.

Tunnel

A faux tunnel may be created by hanging sheets over chairs, or you can purchase a children's plastic tunnel from a department shop at a reasonable price. Don't push your dog into the tunnel if they aren't comfortable doing so. Treats or sticking your head through the other side and calling to them are two other ways to try to entice them.

Make Poles

The weave poles, which require your dog to weave around ten to fifteen vertical poles, are among the most fascinating obstacles to see. You might use PVC pipes driven into the ground or ski poles.

Start with the poles somewhat far apart. To acclimate your dog to the weaving motion, go between the poles while on a leash. Next, let your dog walk between the poles on its own. As your dog starts to get the hang of the motions, you may gradually bring the poles closer together. Here, it's crucial to be flexible, so go slowly to prevent your dog from hurting themselves.

Teeter Board

For many dogs, the teeter board is the toughest barrier since it demands a great deal of trust with moving things.

Start with little items to acclimate your dog to something moving underneath them, such a wobble board, skateboard, or toy wagon. Reward your dog for displaying any curiosity in the item, for placing a paw on it, and finally for managing to balance on top of it. Making it into a pleasant game and creating a favorable connection with moving things are your goals.

When you think your dog is ready, you may construct a homemade teeter board by placing a long piece of wood over a pipe. Once again, you must go cautiously, enticing your dog with goodies and giving them praise every time they come into contact with it and take a few baby steps. Give your dog a prize of goodies for remaining on the teeter when they reach the middle and it starts to move. When they are totally at ease with the action, you may gradually progress to thicker pipes.

Scent Work

Teaching your dog to detect scents doesn't need you to be the owner of a police or military K-9. You don't even need to leave your home, in actuality.

How Does Scent Work?

A sport known as "Scent Work" simulates the job of using detection dogs to find scents, such as explosives or drugs, and alert their handler when the smell has been located. Scent work is a constructive, demanding exercise that gives dogs the chance to make use of their greatest sense in a manner that is enjoyable, captivating, and that develops and reinforces the handler-dog bond.

If you're simply seeking to test it out or teach your dog a new talent, you may follow these instructions in your house. Many dog organizations offer Scent training or "nose work" workshops.

What You Need to Learn Scent Work

Getting the required tools is the first step in training your dog to do Scent Work from home. You can find almost anything you need around the home or buy it on Amazon.

- Essential oil of Birch
- Cut in half cotton swabs
- Tweezers
- a little glass container with a cover
- a cotton swab in a "scent vessel." (To get started, an empty, cleaned mint tin with holes bored in the lid will work.)

- Single-use gloves
- Expensive sweets
- a plastic container with a lid and holes bored in it

Prepare the Scent and Odor Vessel

Put on throwaway gloves and fill each cotton swab with two drops of essential oil in a room separate from where you're working with your dog. Put the cotton swabs with perfume in the glass jar.

Turn the gloves inside out, remove the gloves, wrap them up in a newspaper, and promptly dispose of them outdoors in a garbage bin.

Take a cotton swab out of the jar and place it in the smell vessel using your tweezers.

Put the tweezers into a plastic bag and close it.

If your dog doesn't drool throughout the training session and doesn't become contaminated by food or dirt, you may use the cotton swab again.

Teach Your Dog to Recognize the Fragrance

Hold the treat and the tin with your other hand, spaced approximately a foot apart.

Say "Yes" and give your dog a reward by delivering the food to the hand with the tin when he eventually stops licking or sniffing your treat-holding hand and explores the one with the tin. Note: This is a crucial phase to take. The dog has to be fed where the smell is coming from. You may feed the dog at the tin if it keeps sniffing at it.

Change the tin to the other hand after a few repetitions so the dog isn't dependent on memory to determine which hand to use.

If your dog can accurately recognize each hand's fragrance three times in a row in a matter of seconds, you're good to go.

Help Your Dog Identify the Fragrance

The tin containing the perfumed cotton swab should then be placed into the plastic container.

Use the same procedure again, holding the box in your palm and watching for the dog to make a smell recognition signal. When he does, feed the dog at the box just like you always have.

Once you can do this with ease, set the box between your feet on the ground and carry out the previous procedure once again.

Lastly, you may put the box on the floor while your dog is outside, then bring him inside to see whether he can locate it.

Obedience Training

The majority of folks adore their furry friends. But when your dog isn't educated to avoid bad habits or behave in certain ways, not every moment is fun.

Numerous methods have been handed down from unidentified sources that claim to be the most effective in teaching your dog not to do anything. However, which approach works best and how are these strategies applied?

Find out the most popular dog training methods as well as the ones you shouldn't use.

How Should Your Dog Be Trained?

There are two popular approaches to dog training.

The aversive-based approach is the first. The reward-based approach is the second. Positive punishment and negative reinforcement methods are used in aversive-based (discipline) training with dogs. Reward-based training techniques only provide treats for desired actions in your dog.

In order to educate your dog to behave the way you desire, aversive-based training employs methods that include severe scoldings, physical punishments, and loud, unpleasant sounds. Reward-based training, on the other hand, employs treats every time your dog complies with your requests. Rewarding behaviors with treats, belly rubs, or other activities that please dogs are used to reinforce positive behavior.

Various experts favor different approaches. Whichever one you decide to use is fully up to you.

Some individuals think that training your dog via incentives creates a "event sequence" in which they identify you with happiness when they follow instructions. Conversely, aversive-based techniques make them fearful of you. Your dog follows instructions because of dread of experiencing negative emotions.

Recognize the Learning Process of Your Dog

Like young children, dogs learn a lot. Their IQ is comparable to that of a two-year-old human. Their only concern is the immediate repercussions. They start to comprehend our words as they become older. Certain perceptive breeds may react to up to 250! However, all dogs react more to the tone of the human voice than to the words themselves.

Scientists have classified dog intelligence into three categories:

- Instinctive
- Adaptive
- Working and obedience

Your dog will learn the habits for which they were bred via instinctive learning. The ability of your dog to learn from their surroundings and the world around them in order to solve difficulties is known as adaptive learning. Their ability to learn the duties and orders you teach them depends on their work ethic and compliance.

You should concentrate on teaching your dog using obedience methods and the particular actions you desire from them if you want them to be obedient. Training that is reward- or aversive-based has been shown to be effective. On the other hand, reward-based obedience training is something you should think about if you're teaching your dog to be a loving pet. Your dog won't acquire fear-based reactions thanks to this technique. In reality, it strengthens your affectionate bond with them.

Rewards for Obedience Training

Dogs are intelligent enough to pick up the habits you want. They have the intelligence to figure out what they can get away with as well.

Rewarding a dog with goodies, praise, or love is one of the best ways to teach them to do a certain activity. Above all, the prize that they most want is the finest one to provide them. If they are driven by food, treats could work more than compliments. If they are someone who needs your attention, showing them love might be the finest reward.

Rewarding your dog regularly for desired behavior is the primary thing to concentrate on. Giving rewards for bad behavior is not a smart idea. Your dog should get their reward after they complete the behavior. They become puzzled if you ask them to lay down and then don't reward them with a treat until they get back up. They are not going to know which activity prompted the award.

Manage Repercussions Efficiently

When using reward-based training, your dog must comprehend that misbehaving will have negative repercussions. When individuals behave badly, the implications in this case are that their prize will be withheld.

For example, an elderly person or a little kid may be at risk from a dog that jumps up to welcome them as they arrive inside. If they leap up, do not welcome them or offer them attention in order to teach them not to do so in the future. Once again, you should go out the door and turn around till the dog doesn't leap up. Hold onto a goodie in your hand while you work.

Give the reward to the dog when they don't jump, then carry out the exercise again until your dog doesn't leap up when you enter. This is something you need to attempt with every visitor to your home who makes your dog happy. By doing this, you can be sure that your dog will get a reward for good conduct.

Learning New Capabilities

Keep in mind that your dog has the mental capacity and attention span of a two-year-old while you're teaching them anything new. Your training sessions have to be succinct and direct. Give them no more than fifteen minutes. Keep your attention on a single action or job to prevent confusion.

Verify that the instructions you're using are the same for the desired actions. Your dog could not comprehend if you use the same term again and arrange it in various phrases. Saying "Lie down" in one session and "Fido, lie down or no treat" later in the day can confuse your dog, for example, if you are trying to teach it to lie down. Perhaps they are unsure about what to do.

Basic Training for Obedient Dogs

The American Kennel Club lists five fundamental commands that all dogs need to be able to do. They are as follows:

- Come
- Heel
- Sit
- Stay
- Down

Canine Freestyle

Do you believe you are a dancer? What about your canine companion? Suppose you danced WITH your dog?

Check out Canine Freestyle, a lively canine sport where a dog and handler do a choreographed dance to music, if that seems like a lot of fun.

Canine Freestyle is a fantastic option for a canine activity that enables handlers to express their creativity and goes beyond a predetermined workout regimen. Canine Freestyle Dance, sometimes referred to as "Heel To Music," is a kind of dance routine in which the handler and dog move in time with the music.

The dog never moves from the heel posture for the whole exercise while doing "freestyle heeling." Heel training is mixed with various maneuvers in "musical freestyle" that entail moving the dog away from the handler and the heel position.

The handler may tell the dog when a maneuver is about to be performed by using both verbal signals and body language. The sport honors the beautiful qualities of dogs while fusing the thrilling inventiveness of a dog and handler dancing in time to a selected music.

If you're seeking an innovative method to work with your dog and develop a closer relationship with them, Canine Freestyle is a terrific option for dog sports. It is an excellent option for any type or combination of breeds of dogs. It may be a fantastic method to exercise both you and your dog!

We'd want to get up and dance! What comes next?

You may start teaching your dog at home or hunt for local lessons if you'd want to waltz or tango with Winston or Tucker. It really just comes down to molding habits and then linking them together to teach a dog how to dance.

- Developing fundamental actions such as heeling is the initial stage. Next add more actions such as backing up, weaving between your legs, sitting beautifully, shaking a paw, etc. Every ability has to be taught and developed separately.
- You may ask your dog to do one behavior after another (referred to as "chaining behaviors") after they are proficient in all of these fundamentals as well as a few tricks.

- After you've perfected your routine, play some music and dance to the rhythm of it.
- If there are no classes in your region, you may have to go to one if you want to compete in Canine Freestyle. If so, you may just establish routines from the convenience of your own home.

Fun movements, no stress!

For a number of reasons, some individuals may be too afraid to begin, but don't worry—Canine Freestyle is suitable for every kind of human handler in addition to any breed of dog! Let's talk about a few possible issues:

- What happens if my dog has two...four left paws and I have two left feet?
- Anyone can do Canine Freestyle. Anyone may train their dog certain skills and then dance to music, even the elderly and those in wheelchairs. The objective is to enjoy yourself with your dog, and even if you compete, a significant portion of the score is based on how much fun the team seems to be having. Therefore, instead of worrying about perfection, aim to be entertaining!
- While some competitive teams dress up for events, not all of them do. The main focus of freestyle is on you and your dog enjoying yourself.

Please wait, Your Review is Very Important…

Dear Reader,

I hope this message finds you well. Thank you for choosing to read the Mental Exercise for Dogs Book 2024. Your feedback is incredibly valuable to me, and I would love to hear your thoughts on the book. Whether you've just started, are halfway through, or have finished reading, your perspective matters.

Your feedback is immensely appreciated and will help me enhance future works.

Thank you for taking the time to share your thoughts on Mental Exercise for Dogs Book 2024. Your support means the world to me.

Happy reading!

Howard Briggs

Chapter 4: Building a Routine

Establishing a Daily Schedule

Like humans, dogs like routines and schedules, but each dog is different and has various demands according to breed, age, and temperament. The daily plan given below should be used as a basic reference for your dog, since even dogs of the same age and breed may not need the same things to be happy. For instance, a small puppy might need to go outside more often throughout the day, but an adult dog typically has to go outside for pee breaks every six to eight hours (depending on the individual dog and their requirements). Your lifestyle will also affect your dog's daily routine, although consistency is preferable. We provide you with a useful dog timetable.

General Adult Dog Daily Routine

Morning.

Remember that your dog's morning routine will determine how they feel about the rest of the day, so be sure to provide them with plenty of attention and exercise to get their day started on the right foot. It's better to include them in enjoyable activities, like fetch or jogging. For a change of scenery, you may also try taking them for a stroll in a local park or path.

7:00–7:30 AM: Go potty and wake up

As dogs tend to urinate as soon as they wake up, it is advisable to walk your dog quickly or go on a longer excursion, such a stroll lasting 30 minutes or more. Make sure your dog enjoys and learns from his training sessions. Throughout the day, little lessons will keep students interested and inspired to study.

Morning Short Training Sessions: 7:30–7:45

Training your dog on a regular basis is crucial. Training sessions should be brief and spaced out throughout the dog's day, rather than taking up a whole hour. Remember that training your dog should always be enjoyable for you both! For instance, practice door etiquette with your dog before letting them out for walks and coming home. To stop your dog from rushing out the door, which is quite dangerous, ask them to sit before they enter or depart any entryway.

7:45–8:00 AM: Morning Eats

A Good Dog TimetableEstablish a habit of asking your dog to sit before you put their food dish down so they may consume it. It will instill in them the idea that composure and courteous conduct will lead to rewards like their meal(s). You may even request a longer stay, such as sitting while you make their meal(s). It's also advised to fill their water dish with new water every day.

8:30–9:00 AM: Restroom

To avoid injuries like bloating, let your dog at least thirty minutes to calm their stomach before letting them go outside for a pee break. Try taking your dog for a little walk to help stimulate their digestive system if you notice they are having problems going pee. Make sure to give your dog praise and treats when they successfully relieve themselves outdoors.

Crate, 9:00 AM–1:00 PM

You may cage your dog if you have to leave during these hours to go to work (this only applies if your dog is housebroken). Have your dog practice going into their cage with the assurance that it is their safe haven by using the verbal cue put or "Go to your crate/bed." When your dog goes inside their cage, it will teach them how to relax, be independent, settle down, and even take a nap! Give them tasty chew toys to occupy their time, such a Kong packed with their favorite candies.

Midday

To provide your dog with attention and a pee break if you are unable to be at home during this time, think about hiring a pet sitter or dog walker. It's crucial to make sure your dog receives some social connection throughout the day. It fosters their social skills development and confidence-building. You may set up playdates with other canine friends or take them to a nearby dog park. Enrolling them in a training program, where they may socialize with other dogs and pick up new abilities, is also a smart choice.

1:00–1:30 PM: Restroom

Taking your dog outside for a midday bathroom break will be enjoyable for them. If you are unable to take your dog for a walk after work, you may want to think about getting a dog sitter or walker to assist you.

1:30–5:30 P.M. Play, Instruction, and Socialization

Good Dog TimetableDogs are naturally playful! During these hours, make an effort to schedule some time for your dog to play with you or interact with other dogs. Play games like retrieve the ball or with a toy to pass the time while you and your dog are together. Remember that you can squeeze in some training time while playing. You may ask your dog to sit, for instance, release the sit, and give him a toy as a treat. Another example of how to stop resource guarding is to teach your dog the command to give back. Socialization is an important aspect of your dog's life; take them on play dates to allow them to interact with other dogs that they get along with. You will always want your dog to be confident and friendly with other dogs. You may arrange to get together with the pets of your family, friends, or even your neighbor.

Once again, if you are unable to provide your dog with playtime, you may want to think about getting a dog sitter or enrolling your dog in a dog daycare facility, where they will be able to interact and play with other dogs. Practice recalling the come cue, often known as playtime.

Evening

5:30–7:30 PM: Walk/Potty Break

Take your dog for a lengthy walk or a toilet break. For a longer trek, the bathroom break and stroll might last up to thirty minutes. Walking your dog is a great opportunity to teach them good leash etiquette! Walking your dog will become more joyful for both of you instead of your dog tugging you while wearing a leash.

6:30–7:00 p.m. Take A Seat

I think you should use this time for yourself! Your dog may be left in their crate for a short while, or you can send them to their bed.

7:15–7:15 PM: Dinner

This is the perfect moment to serve your dog supper, or you may wait until it's most convenient for both of you. The most crucial element is to make an effort to time things consistently.

7:45–9:30 PM: Play or unwind

Watching your favorite TV program while lounging with your dog will be a soothing experience!

Make the most of this opportunity to show your dog some affection by stroking, snuggling, or even doing some handling exercises. You may play with your dog at this time as well.

9:30–10:00 PM: Final Potty Break/Sleep Schedule for Good DogsAfter giving your dog one more potty break, put them to sleep. You can be certain that your dog would be ready for a sound slumber after an active, long day!

Although caring for a dog might sometimes seem like a full-time job, it will keep your dog happy, healthy, and active! Your relationship and affection for your dog will grow even closer as a result of all of your enjoyable activities together.

Establish your own healthy, good-dog plan and make sure your dog has a regular routine.

Monitoring Progress

You and your mentors will find it very helpful to have a record of your dog's development. Frequently, an issue arises from the combined impact of several training sessions and may only be identified via documentation. A handler has several options for monitoring training. These are just a few of the many I've used.

Logs on paper
They are simple to operate and portable. The log and a pen are all you need.

Journal
A cheap lined notebook is the easiest. Just record everything that occurred throughout each training session.

Advantages:

Simple to use, portable, unformatted, and unrestricted writing

Cons:

weather-related unformatted tough to recall what is essential to write down; challenging to communicate with others; difficult to measure progress and identify the relevant information in the text
Poor handwriting is difficult to defend.
Formatted Log Sheets

To design a training form, use any word processor. All you have to do is print them off, pierce them, and place them in a binder. Focus on utilizing check boxes and similar tools to make the most critical information visible at a glance.

Advantages:

Simple to scan for crucial data
standardized—each team member uses the same form

Cons:

restricts space for lengthy notes
Drive Record

While I still work on the drive with my operational dog, they are most helpful with younger dogs. Observe how you can see the information that matters.

Lookup Records

With a form like this, everything you need to monitor is visible. KISS stands for "Keep it Simple, Stupid." Make everything visually appealing and simple to swiftly skim for information. Identifying trends and issues is the goal here.

On the other hand, I jot down details to make sure I remember everything and can quickly review the specifics of that day. Despite the fact that these logs were written about two years ago, my insightful remarks fully evoke the experience.

To be included are the following categories:

Specifics of the search operation. Describe the precise actions the dog took and how the issue came to be.
Remarks from your teammates or mentor: they will notice things that you miss. Most dog issues are the result of actions taken by the handler. You often act without recognizing it!
The handler's remarks: What did you witness?
Items to improve on: What guidance did you get from others? Next time, how will you organize the training?

Just keep in mind to bring your logs to every training session.

Computer Records

We are now able to carry papers with us wherever we go thanks to smartphones, iPads, and tablets. With a single button click, they also enable us to send that information to others and our home computer.

Advantages:

Continually adjustable and backed up
the capacity to distribute and connect pictures and videos
searchable, date-stamped, and free from handwritten errors

Cons:

requires a full charge and is not as simple to use in the field
less simple to standardize Everybody uses a different device, and not every device can be utilized by every device. cross-platform issues (software used by iPhones differs from that of Androids and PCs)
privacy issues: don't forget to backup your data across many devices.
I've converted to computer logs since I always have my phone with me and I can see videos from that day's instruction to go along with my daily journal. I've seen trainers use word processors and excel.

Evernote is a great app that works on almost every platform. Every note may include forms, documents, videos, online links, images, and more, and it can be accessed from any location. Almost everything has been transferred here.

Everything on my phone is backed up to the "cloud" via Evernote. I can use any device to view it, and I have several methods to share each individual message with others. It is password secured.

For K9 search and rescue logs, use Evernote.

My only gripe is that training records become as hard to read as a personal journal until I can figure out how to make a template for them. I just copied the aforementioned paper training records and created a template. I highlight the term and tell the computer to emphasize it rather than circle each alternative.

Advantages:

Videos, documents, online links, schematics, drawings, and almost anything else may be embedded into each note, making it endlessly shareable across all platforms. Each note can be quickly backed up and organized using keywords, tags, and notebooks.
date-stamped and searchable; moreover, it is password-protected.

Cons:

unformatted: You'll need to either learn how to make a template that makes material simple to scan, which is a little learning curve but takes longer than writing by hand. Be cautious about what you give.

Chapter 5: Mental Exercise and Behavior

Reducing Anxiety and Stress

1. Take Your Dog for Exercise

The apparent solution to your dog's separation anxiety is to never leave them alone. Since most pet owners cannot afford that, exercising your pet to exhaust them and foster a relationship with them is often a simple solution!

It might be beneficial to take your dog for a long walk or game of fetch before you leave, since anxiousness can lead to hyperactivity. Talking to them and making plenty of physical contact with them are also helpful at this period. Additionally, much as with humans, exercise helps reduce stress by releasing feel-good endorphins.

2. Touching

A dog that is nervous usually finds solace in nothing other than the touch of its owner. When you see your dog exhibiting symptoms of anxiety, try to spot them early on and provide them with plenty of attention, such as picking them up or snuggling with them on the sofa.

3. A massage

A massage is known to soothe and relax even the most tense person, but did you know that it also has a similar calming effect on dogs? Muscle tensing brought on by anxiety may be relieved in part by massage treatment. Using lengthy strokes, begin at the neck and work your way down. While the other hand is massaging the dog, try to maintain one hand on it. With time, you may even be able to pinpoint the exact spot on your dog where they store their tension and focus just on it.

4. Musical Rehabilitation

It has been shown that music therapy is advantageous for us as well as our furry and canine companions. Whether you're at home, in the vehicle, or anywhere other than with your pet, music has the ability to soothe and relax. By obstructing the street or frightful sounds that some dogs find upsetting and causing anxiety, music may also help dogs who are sensitive to noise.

5. The Time-Out

While nervousness in and of itself is not a harmful behavior, it might be helpful to put your dog in a time-out when they exhibit erratic behavior. Your pet may become less agitated if you isolate them in a secure area. Perhaps there's low lighting, some extremely quiet music playing, and/or aromatherapy accessible in that area (see "Alternative Therapies" below).

You might also consider using a ZenCrate as your pet's general escape pod and for time-outs. The purpose of the ZenCrate was to assist dogs suffering from a range of anxiety disorders. It resembles a regular crate but offers decreased light, comfort, security, and noise cancellation (via sound insulation) in addition to vibration isolation. When your dog enters, a motion-activated sensor activates a quiet fan that lets in fresh air and helps block out noise. The container may be pre-programmed to play music. With its detachable door, your dog may feel comfortable and come in at any moment.

6. Calm T-shirts and coats

Like a swaddling wrap on a newborn, calming jackets and t-shirts envelop a dog's midsection with gentle, continuous pressure. For dogs who experience anxiety related to travel, separation, noise, or strangers, it is advised.

You may select from a variety of brands and models based on your dog's size. Check out the Comfort Zone Calming Vest, American Kennel Club Stress Relief Coat, and ThunderShirt Anxiety Jacket.

7. Complementary Medicine

The items described below are non-invasive and will not damage dogs experiencing anxiety, despite the paucity of data supporting the benefits of alternative treatments. These are treatments that work better when coupled with the ones mentioned above or used on their own. Prior to starting any alternative therapy, make sure you do your homework and speak with your veterinarian.

Preventing Destructive Behavior

Although dogs are devoted, loving members of our homes, sometimes their behavior may become destructive, resulting in damage to personal property and household things. Although this kind of behavior often irritates people, it may be more harmful to your dog's health.

There are several explanations for why a dog may act destructively. To assist explain these troubling behaviors, we will examine the causes of destructive dog behavior in this article.

Dog behavior examples that are damaging

There are many ways that a dog might act destructively. Dogs that exhibit destructive actions most often include:

1. Chewing
2. Digging
3. Barking
4. Jumping
5. Other strange behaviors can have a damaging quality.

For dog owners, any of these may provide challenges. Chewing degrades things, digging damages your yard and floor, barking annoys neighbors, and so forth. To stop these behaviors from occurring, it is necessary to comprehend why they start in the first place. Your dog may seem to be a violent ball of fur with the intention of destroying everything in its path, but these behaviors may have more straightforward causes than you may realize.

What leads to harmful conduct?

There are several explanations for why a dog may be acting destructively.

One of the simplest explanations might be that pups are teething and want to chew anything in their immediate environment to ease their pain. Even while this tendency usually ends as permanent teeth erupt, it could carry over into adulthood. Redirecting your dog to something suitable for chewing, such a chew toy, is the best course of action since chewing encourages exploration of their surroundings.

The causes of destructive behavior in adult dogs differ. Dogs who behave destructively most often have one or more of the following reasons: boredom, fear, attention seeking, medical issues, and separation anxiety. The most common causes of these issues are insufficient exercise, mental stimulation, or training. In extreme situations, the behavior has a much more complicated cause, such as traumatizing events, phobias, or genetics and family history.

Dogs don't just decide to choose aggression and start destroying their surroundings, regardless of the reason behind it. Dogs act destructively as a means of letting off steam or as a way to cope with stress and anxiety. Dogs that are nervous will respond by chewing, leaping, or barking, much as humans who are nervous would exercise or chew gum.

How to put an end to your dog's destructive habits

Understanding why a dog is behaving in a certain manner is the first step towards stopping harmful canine behavior. Finding the reason facilitates situation assessment and facilitates the process of issue solving.

It is essential to intervene before harmful conduct develops into a habit. The following are some useful pointers for putting an end to and avoiding damaging dog behavior:

Dog proof your house by hiding any valuables you believe should be kept out of your dog's reach. This will make it less likely that priceless objects will be harmed and provide you with a clearer environment in which to control the harmful behavior.

Watch your dog closely: Make sure your dog isn't acting destructively while you're at home by keeping an eye on them. By keeping an eye on the dog, you can also teach it acceptable behavior by leading and redirecting it.

Give your dog a chew toy: it's normal for dogs to chew. Providing children with secure chew toys helps steer them in the correct path. What they can and cannot chew is taught. They need to limit their chew toy possessions to one or two at a time. If there are too many toys on the floor, they can assume everything is fair game and begin gnawing on things they shouldn't.

Give your dog enough exercise; a weary dog is a happy dog. When dogs don't receive enough exercise, they may grow agitated, and that's usually when the destructive behaviors start. A weary dog will be less likely to wreck things and will choose to rest and unwind while they are at home.

Put your dog to the test: Mental stimulation is just as vital as physical exercise for dogs. Even more so than physical exertion, mental stimulation helps them learn, reduce stress levels, and becomes fatigued fast. Playtime with other dogs, training sessions, scent work, socializing, and interactive toys are all excellent ways to stimulate the mind. These occupy their thoughts and lessen the want to behave destructively.

Train your dog: Consistent training sessions can provide dogs the cerebral stimulation they need while also instilling in them a sense of discipline by learning right from wrong. All types of training, from agility to obedience, may lessen negative behavior and win you respect, which will make it simpler to stop undesired actions.

Finally, refrain from using punishment to get your dog to behave differently. If your dog is already seeking a way to release his or her frustrations or anxieties, punishing them will only make their destructive conduct worse and may even make the problem worse.

Enhancing Social Skills

During the first few months of your dog's existence, social skills training is an important aspect of training. It is essential to their wellbeing and to their ability to interact with humans and other canines without stress.

Starting socialization with your dog while they are still a puppy is preferable since it is a lifetime process. Dogs may always pick up new tricks and develop their social abilities, regardless of their age. Therefore, it is never too late to start.

When to Begin

Socialization of dogs should start as soon as possible. The best period for pups is from three to fourteen weeks of age.

This is a procedure that a reputable breeder ought to have begun for you before your puppy was even born. Since the first three months of a dog's existence are crucial, exposing your puppy to as many people, pets, and situations as you can will help them have a wonderful start in life. Always be sure to accomplish this in a methodical and constructive manner.

Promote Good Relations

Your dog should enjoy every new encounter that they have.

Steer clear of circumstances that might be intimidating or overwhelming for your dog if you want to reduce the likelihood of them being anxious. Later on, your dog can develop fear or hostility as a result of this. This is crucial for older dogs since there may be triggers already. Having delicious goodies on hand and providing plenty of positive reinforcement can help keep them calm and interested.

Expose Yourself to Varying Environments

Visit as many different locations as you can with your dog.

You may socialize your dog in a variety of settings, such as parks, beaches, pet-friendly shops, and friends' houses. They will be exposed to a variety of sights, sounds, and scents as a result. They will pick up more social skills and have more confidence in unfamiliar situations the more they experience. The ideal approach is to expose gradually. A socializing checklist can be useful.

Get Along with Other Dogs

Set up a playdate for your dog with other well-mannered canines.

To help your dog get used to encountering animals of different shapes and sizes, try to obtain a mixed breed. Joining a dog walking club or putting your pet in dog day care are easy ways to do this.

Classes for Dog Training

Bring your adult or puppy dog to a nearby dog training program.

These sessions help teach your dog obedience skills and provide organized socializing chances. Positive reinforcement strategies are used in holistic dog training sessions, which makes them excellent.

Get Along with a Variety of People

Introduce all kinds of people to your dog. individuals of varying ages, sexes, and racial backgrounds.

Introducing them to a variety of individuals can enable them to stop being afraid of strangers they encounter while out for walks or at the veterinarian or dog groomer. Encourage both your dog and visitors to remain composed when they first meet. Once again, utilize rewards and compliments to promote pleasant encounters.

Speak With The Professionals

Speaking with professionals is the best course of action if your dog is older and aggressive.

Consult a professional dog trainer if your dog exhibits fear or is susceptible to certain stimuli, such as thunderstorms or other dogs. They will instruct you on how to counter condition and desensitize your dog to these stimuli. To assist in overcoming behavioral problems or challenges you may be facing, professional dog trainers or animal behaviorists may provide advice and training programs.

Acquire Knowledge Of Dog

Gaining proficiency in dog language will enable you to read your dog's body language.

It's critical to understand the early warning indicators of fear, anxiety, aggressiveness, and playfulness in dogs. With this understanding, you'll be able to intervene when necessary and promote constructive relationships.

Regularity

When it comes to teaching and socializing your dog, be dependable.

It takes consistent practice and exposure to a variety of environments to develop and preserve those critical social skills.

Be patient.

Like people, every dog is unique and develops at their own rate. Keep your cool and enjoy the little victories along the road. Taking your time will help your dog acquire exceptional social skills and have a happy, balanced existence. Socialization is a continuous process.

Strengthening the Human-Dog Bond

A person's best buddy is a dog! That doesn't imply, however, that you can't show your dog your affection in other ways.

Instruction

Not only can teaching your dog new tricks improve their manners and conduct, but it also strengthens your bond with them and is a psychologically stimulating hobby. Your dog must be tuned in to you throughout training to receive direction. Their confidence will grow as they learn new orders and get goodies and praise from you!

It's time for them to learn basic instructions if they don't already know them! If your dog is more experienced, you may teach them fancy new tricks or engage them in mental exercises like agility or nose work.

Work Out Together

We can always benefit from exercise, and our pets do too. Walking or running beside your dog may be an excellent way to strengthen your relationship.

Additionally, make sure the activity you select is appropriate for the breed and activity level of your dog. Running behind you as you ride might be beneficial for breeds that have high energy and can manage endurance. A pleasant walk might help if your dog is overweight, elderly, or of a short nose breed.

grooming

When your dog is properly socialized, treating them to a spa day may become a treasured hobby. Your dog may not like a bath, brush, or nail trim right now, but if you are kind to them and give them goodies, they will eventually learn to like the attention.

Engage in Play Together

Discover what your dog enjoys doing most, and then give it your all. Engage your dog in active games like tug-of-war or fetch, and your joy will be infectious to your dog. Furthermore, engaging in active play might be more exciting than just doing regular exercise.

Spend Time Cuddling and Petting

Throughout the day, give your dog a short pat on the head every now and then. Make sure, nevertheless, that you give your pet some deliberate time for physical interaction. For example, if your dog loves to cuddle, get down on the floor with them and spend some quality time together.

Pet intentionally, taking the time to learn about your dog's preferences. Find out which of their favorite snuggle times they like most, and spend some time together giving each other belly rubs or ear scratches.

Regular, Regular, Regular

Dogs do best with regular routines. Together, you may establish a routine so that they will know what to anticipate from you and when, which will only increase their faith in you. They'll also pick up on expectations and meet them, and dogs, as we all know, like making their owners proud!

Additionally, communicate with them consistently. To avoid confusion, stick to cue words they are already familiar with. Our pets are curious about what we expect from them.

Chapter 6: Nutrition and Mental Exercise

The Role of Diet in Cognitive Health

Have you ever wondered how your pet's nutrition may affect their learning and cognitive capacities? Let me just say that it's really intriguing! The way your dog's brain functions and general mental health is affected may be greatly influenced by the kind of food you feed them. Making the appropriate food choices for your dog's diet is essential for their cognitive growth since, like people, dogs need the correct nutrients to maintain the health of their brains.

Now let's talk about the What, Why, and How of cognitive care for dogs! Admittedly, we tend to forget how critical it is to enhance dogs' cognitive health, especially when we see changes in their behavior or learning capacities.

First, let's define "what" cognitive health is.

Cognitive health refers to our brain's capacity to carry out many tasks effectively, including:

Capacity to reason; Learning; Recall; Motor skills; Sensation-based reactions

Why Using a Diet to Help Your Dog's Cognitive Health

Enhance the Cognitive Health of Your Dog with DietFor senior dogs, why is canine cognitive health so important? Our goal is to provide our closest friends the longest and happiest life possible. A dog who is unable to engage in its normal activities is quite depressing. As canines age, cognitive impairment may become a problem. It is not enjoyable to see your closest friend suffer, particularly if you are aware of the potential for dissatisfaction. For brains to perform at their peak, the correct fuel is required. The key is a high-quality diet full of nutrients that support the brain.

Remember that cognitive health is crucial for a puppy's brain growth and well-being, not only for aging adults. greater attention, quicker learning, and greater memory retention are all facilitated by a healthy brain and are necessary for our dogs' effective training as well as their long-term survival.

Why is eating so crucial to your dog's cognitive health, especially in younger canines?

Maintaining your dog's cognitive health and improving their learning potential need a balanced, nutritious diet. Dogs need a range of nutrients, such as vitamins, antioxidants, and omega-3 fatty acids, to keep their minds functioning properly, just as people do. These nutrients are essential for maintaining dogs' cognitive abilities and ability to retain memories. You can keep your dog attentive, engaged, and ready to learn new instructions and tricks by feeding them a healthy diet. Thus, the next time you're selecting food for your dog, remember to take their cognitive skills into account. After all, maintaining a balanced diet is important for your dog's mental clarity and general cognitive health in addition to their physical well-being.

Why it's crucial to improve your dog's food for cognitive health in allergy-prone canines

Dogs may have food sensitivities and allergies, much like humans, which can impair their cognitive abilities. The food that you feed your pet may have a profound impact on both their general health and ability to learn. You can make sure they remain focused, alert, and prepared to take on any new skills or training difficulties by feeding them a well-balanced and nutrient-rich diet. Therefore, let us ensure that our dogs are receiving the greatest nourishment available for their cognitive well-being!

How to Use Nutrition to Help Your Dog's Cognitive Health

The ways in which diet may enhance your dog's cognitive health are astounding. A little food adjustment may have such a big impact on our pets' life! Canine cognitive dysfunction (CCD) is the term used to describe the decline in a dog's neurological and cognitive functioning. In humans, CCD is referred to as dementia. Dogs' declining cognitive health and everyday performance are mostly caused by environmental and lifestyle variables.

As they say, "you are what you eat," and dogs are no exception! Our dogs will never get a proper quantity of nutrition unless pet owners realize that giving them a bowl full of highly processed, high-carb food that is packed with by-products and fillers is not healthy. Long-term issues including cognitive impairment, heart disease, obesity, diabetes, and cancer may be brought on by this kind of diet. The answer to enhancing cognitive health is nutrition. The nutritional value of healthy meals is essential for treating chronic oxidative stress, chronic inflammation, docosahexaenoic acid (DHA) insufficiency, and cognitive impairment in dogs. nutritional treatments may be used to tackle all of them.

The consumption of alternative foods may greatly lessen harm to the body and brain. For instance, coconut oil has over 15% MCT oil, an alternative ketone that provides energy to the body and enhances learning, executive function, and memory.

Canine Immune Support from MushroomsWhat foods can you feed your dog to improve their cognitive health via diet?

- Due to its high MCT content, coconut oil is excellent for enhancing memory, reflexes, and sensory responses.
- Salmon is high in omega 3, as is salmon oil. Omega 3 improves general function while safeguarding the brain!
- There are many antioxidants in cranberries! Your cells are protected by antioxidants against free radicals, which may lead to a number of illnesses.
- Peas are a great source of antioxidants, vitamin C, and E.
- In addition to keeping the eyes bright, carrots lower brain inflation. Lutein, found in carrots, has anti-inflammatory, anti-cancer, antioxidant, and neuroprotective properties.
- Lion's Head Mushrooms may enhance cognitive and memory performance.

- The secret to improving your dog's cognitive health with diet is to start early! Don't wait for your buddy to become too elderly to begin adopting a healthier diet. It's essential to offer your puppy the greatest nourishment possible to ensure their healthy growth and prevent illness. Select a diet that

> → Is high with antioxidant
> → Contains DHA
> → B vitamins
> → Omega-3 fatty acids
> → Mushrooms

The greatest strategy to improve your dog's cognitive health via diet is always prevention. You will enjoy your companion for many years to come if you make sure your puppy is getting what it needs! Remain safe and fierce!

Supplements for Brain Health

Is it time for you to think about getting your dog some cognitive supplements? Upon the elegant ascent into old age of man's best companion, behavioral, memory, and learning capacities may alter. These changes that older canines go through may be compared to what people with diseases like Alzheimer's go through. Known as Canine Cognitive Dysfunction (CCD), 14–35% of canines are thought to be affected by this degenerative illness. The quality of life of our dogs is compromised by cognitive decline, which also strains the bond between pet and owner. Thankfully, research into canine health has produced a number of cognitive-enhancing brain supplements for elderly dogs, with the goal of maintaining their mental sharpness, delaying cognitive deterioration, and guaranteeing a higher standard of living.

But which dietary supplements work well and are reliable? You may get all the information you want to make an educated decision on cognitive and brain supplements for dogs by reading this guide. You'll discover all the information you need, including aspects to take into account when selecting a supplement.

Additionally, we've put up a list of the top canine cognitive vitamins to help make the decision less intimidating. Let's first discuss the symptoms of canine cognitive impairment.

What Indicates a Dog May Be Experiencing Cognitive Decline?

Canine Cognitive Dysfunction (CCD), another name for cognitive loss in dogs, is comparable to Alzheimer's disease in people. It may show itself as a variety of behavioral changes, some of which may start off little but eventually become more obvious. The following are important indicators that your dog could be going through cognitive decline:

1. Disorientation: Even in familiar surroundings, your dog may seem bewildered or disoriented. They may get trapped behind furniture, in corners, or lose track of where doors are.
2. Sleep Pattern Changes: Dogs with CCD may sleep less at night and more during the day, which might cause restlessness or an increase in activity at night.
3. Loss of House Training: They can begin to have "accidents" inside, lose track of time when they went outside, or forget where the door is.
4. Changes in Interaction: Dogs may become clingy or lose interest in greetings, stroking, and social interactions. It might seem that they forget familiar faces, objects, or locations.
5. Shifts in Activity Levels: They can become less interested in playing, exploring their surroundings, or engaging in family activities.
6. Anxiety and Enhanced Irritability: Elevated anxiety, specifically related to separation anxiety, or heightened irritability may indicate a deterioration in cognitive function. Alterations might include heightened hostility or the emergence of new phobias and worries.
7. Modified Appetite: Modifications in eating patterns, such as cutting down on meals or seeming to skip meals, may be a sign of cognitive issues.
8. Pacing, circling, or looking at walls are examples of repetitive or compulsive behaviors that may indicate cognitive deterioration. Some dogs could get obsessed with a certain thing or toy.

9. Forgetting: Although dogs have excellent hearing, they sometimes forget orders they have heard for years or act as if they are not hearing you.

It's critical to seek veterinarian care if your dog exhibits one or more of these symptoms. These symptoms may be indicators of cognitive loss, but they may also be markers of other more significant health problems. An early diagnosis may result in a treatment plan that can greatly enhance your dog's quality of life and may include medication, food modifications, lifestyle changes, and supplements that promote cognitive function.

What Constituents Should I Search for in an Elderly Dog's Cognitive/Brain Supplement?

Some essential elements in senior dog brain or cognitive supplements have been shown in studies to support brain health and cognitive performance. Here are some points to consider:

1. Omega-3 Fatty Acids: These vital fatty acids, in particular DHA (docosahexaenoic acid) and EPA (eicosapentaenoic acid), boost brain function and health, lower inflammation, and improve cardiovascular health. Fish and flaxseed oils contain these fatty acids.
2. Phospholipid phosphatidylserine is essential for the structure and operation of cells, especially those in the brain. It has been shown to enhance dogs' stress reaction, memory, and cognitive performance.
3. Antioxidants: Oxidative damage in brain cells may be fought off by vitamins E, C, and other antioxidants, which may halt cognitive deterioration. Additionally beneficial to general health and the immune system are antioxidants.
4. Ginkgo Biloba: Known for its neuroprotective properties, this age-old herb may assist in enhancing canine memory and cognitive performance, just as it does for people.
5. B vitamins: In particular, vitamins B6, B9, and B12, assist the production of neurotransmitters, promote brain health and function, and may lower the risk of heart disease.

6. L-theanine: An amino acid that is present in green tea, L-theanine helps dogs relax and behave less anxious and stressed out without making them sleepy.
7. The body uses SAM-e, or S-adenosylmethionine, to aid in cellular repair, especially in the liver and brain. It may help elevate mood and cognitive function.
8. Resveratrol: Rich in antioxidant and anti-inflammatory qualities, resveratrol is found in grape skin. Preliminary study on brain health is promising, but additional research is required.
9. Medium Chain Triglycerides (MCTs): Found in palm and coconut oils, MCTs have been shown to enhance brain energy metabolism and reduce the accumulation of amyloid proteins linked to dementia. MCTs may provide brain cells an alternate energy source.

Recall that you should always speak with your veterinarian before beginning a new supplement regimen to make sure the supplements you choose complement your dog's overall health routine. Your veterinarian can help you choose a product that has undergone extensive testing and quality control to ensure its effectiveness and safety.

What Other Elements Need to Be Taken Into Account When Selecting a Cognitive Supplement for Dogs?

It's important to give serious thought to the safety, effectiveness, and suitability of cognitive supplements for dogs. Think about the following:

1. Particular Requirement or Situation: Ascertain your motivation for considering a supplement. Is it for anxiety, general brain health, age-related cognitive loss, or any other particular condition?
2. Ingredients:
   - ➔ *Safety: Verify that every item is suitable for ingestion by dogs. Steer clear of supplements that include dubious ingredients, artificial coloring, or hazardous fillers.*
   - ➔ *Efficacy: Look for components like omega-3 fatty acids (DHA and EPA), antioxidants like vitamin E and C, L-carnitine,*

alpha-lipoic acid, and phosphatidylserine that are known to promote canine cognitive function.

➔ *Natural vs. Artificial: Certain pet owners like organic supplement sources, such herbs like gotu kola or ginkgo biloba.*

3. Evidence of Effectiveness: Look for items that have been clinically or scientifically shown to be effective in promoting canine cognitive function.
4. Brand Reputation: Choose trustworthy businesses with a solid track record in the pet care sector and positive customer feedback.
5. Possible Interactions: To prevent any interactions, take into account any other nutrients or drugs your dog may be taking.
6. Certifications: Verify if the product has received certification from a third party, such as the National Animal Supplement Council (NASC) or other such associations. Higher manufacturing standards may be indicated by such certifications.
7. Age and Size Appropriateness: Verify that the supplement is suitable for the size, breed, and age of your dog. Needs and dosages might change.
8. Administration Ease: Supplements may be taken as liquids, chews, or pills. Select a form that your dog can readily eat.
9. Potential Side Effects: Learn about and comprehend any possible negative effects that the supplement may have. Once your dog starts taking the supplement, watch him to make sure he has no negative effects.

Chapter 7: The Importance of Rest and Relaxation

Relaxation Techniques for Dogs

Our pets are treasured family members that provide us with unwavering affection and company. But our animal pets may feel stress and worry, just as people do. Pets who are stressed out might be caused by thunderstorms, separation anxiety, or strange surroundings; it is important to know and use soothing strategies to help them unwind. We will look at practical ways to calm our agitated dogs and create a calm atmosphere in this chapter.

Comprehending Pet Stress: First and foremost, it's essential to identify the telltale symptoms of stress in our dogs. Excessive barking, panting, pacing, shaking, destructive behavior, hunger loss, and even hostility are typical signs. Pets may also show physical signs including dilated pupils, heightened stress hormone levels, and an accelerated heart rate. Furthermore, certain animals may retreat and hide in remote locations.

Establishing a peaceful atmosphere: It's important to provide a peaceful atmosphere for your pet in addition to being aware of the telltale indications of stress. Moreover, this may be accomplished by offering your pet a secure haven where they can hide when they're feeling overwhelmed. This might be a special spot in your house with their bed and favorite toys, or it could be a quaint nook. Therefore, it is crucial to make sure that this area is free of loud noises and other distractions that might cause tension.

Additionally, aromatherapy and relaxing music may be quite effective in calming stressed-out dogs. Their heart rate may be lowered and relaxation can be encouraged by listening to slow-paced, gentle music. Moreover, certain smells, such as chamomile or lavender, might help animals relax. You may sprinkle their allocated area with diluted aromatic mists or use an essential oil diffuser that is suitable for pets. Additionally, goods made especially for dogs, such pheromone diffusers, might help lessen anxiety.

Massage and Touch Therapy: In addition, dogs may profit from touch therapy in the same way that people do. Your pet might benefit from a gentle massage to help them relax and release stress. It also lessens muscular stiffness and enhances circulation. Additionally, you may use a variety of methods, like circular movements or lengthy strokes, based on your pet's preferences. To make sure they are at ease and enjoying the experience, keep in mind to be kind and pay attention to their reaction.

Exercise and Playtime: Moreover, consistent playtime and exercise are essential for lowering pet stress levels. Exercise encourages the release of endorphins, or "feel-good" chemicals, and aids in the release of stored up energy. Playing interactively with your pet also improves your relationship and gives them cerebral stimulation. In addition, engaging in activities like puzzle toys or toys that release treats helps keep them busy and divert their attention from stressful situations.

Pets, on the other hand, benefit greatly from regularity and predictability. Creating a regular schedule for eating, playing, and exercising each day might help them feel less anxious. Keeping a regular routine also gives people a feeling of security and stability. Furthermore, since they know what to anticipate, maintaining a schedule during stressful events like thunderstorms or fireworks might help them feel less stressed.

expert Assistance: In spite of our best efforts, there are situations when our dogs need expert assistance to manage their stress and anxiety. Additionally, speaking with a licensed animal behaviorist or veterinarian may provide insightful advice catered to the unique requirements of your pet. To assist your pet in overcoming their triggers and concerns, they could also suggest methods like behavior modification exercises or desensitization. If required, they may also recommend the proper prescription or vitamins.

Please wait, Your Review is Very Important…

Dear Reader,

I hope this message finds you well. Thank you for choosing to read the Mental Exercise for Dogs Book 2024. Your feedback is incredibly valuable to me, and I would love to hear your thoughts on the book. Whether you've just started, are halfway through, or have finished reading, your perspective matters.

Your feedback is immensely appreciated and will help me enhance future works.

Thank you for taking the time to share your thoughts on Mental Exercise for Dogs Book 2024. Your support means the world to me.

Happy reading!

Howard Briggs

Conclusion

In the constantly changing field of canine mental stimulation, it's important to take advice from professionals and draw motivation from actual success stories. Your quest to realize your dog's potential may be aided by these success stories and the guidance of experts in the field.

Now, let's explore some expert advice on how to improve your dog's quality of life and strengthen your relationship with your four-legged companion.

- Utilize puzzle toys: These are a delicious method to treat your dog and keep their brains active. They are available at different degrees of difficulty, so you can provide your dog with the ideal challenge.
- Rotate toys: Over time, dogs may become bored with the same toys. Toys should be rotated in and out of usage to keep things new. This will maintain your dog's interest and enthusiasm for playing.
- Learn some new tricks: Teaching your dog new skills is entertaining for both of you and helps your dog's mind. To motivate your dog, break the trick down into manageable stages and give them praise.
- Take advantage of the extraordinary sense of smell that dogs possess by going on scent walks. To provide your dog mental stimulation and enrichment, encourage them to investigate new scents and settings.
- Make a treasure hunt for your dog by hiding toys or goodies around your yard or house and letting him explore for them. Their ability to solve problems will be tested by this exercise, which will also be entertaining and interesting.
- Dogs are gregarious creatures, so socializing with other canines is a terrific way to keep your mind active. Plan playdates with other dogs or take your dog to a dog park nearby to help them socialize and have fun.
- Play dog sports: Agility, flyball, and dock diving are excellent forms of mental and physical exercise for your canine companion. As you collaborate as a team throughout these activities, your relationship with your dog may also develop stronger.

- Use interactive feeders: Try feeding your dog from an interactive feeder rather than a regular dish. By making your dog struggle for their food, these gadgets slow down their eating and stimulate their mind.
- Try giving your dog a massage: Giving your dog a massage is a terrific method to provide them mental stimulation and relaxation. It may enhance circulation, lessen tension, and improve your relationship with your dog.
- Have a game of hide-and-seek: Hide-and-seek is a timeless game that is a great method to stimulate your dog's thinking and problem-solving abilities. Call your dog to locate you by hiding in various spots around your house or yard.

After discussing some professional advice, let's look at some actual success stories to motivate you to help your dog reach his or her greatest potential.

The Max story: Once a high-energy dog with anxiety and destructive tendencies, Max was a border collie. His owner started to include mentally stimulating things in his everyday routine, like agility training, scent walks, and puzzle toys. As a consequence, Max had less anxiety and developed into a happier, more well-mannered dog.

The story of Molly Rescued from a difficult life, Molly was fearful of other dogs and aggressive against them. Her owner began enrolling her in dog training programs and participating in dog sports like flyball. As Molly gained greater self-assurance and ease with other canines, she developed into a gregarious and well-rounded buddy.

Charlie's journey: Charlie, an elderly dog, started exhibiting symptoms of cognitive impairment, such as disorientation and bewilderment. Activities for cerebral stimulation, like interactive feeds and smell walks, were included into his daily routine by his owner. Through these pursuits, Charlie was able to enjoy a longer, more fulfilling life throughout his senior years and slowed down the rate at which his cognitive loss was progressing.

These tales show how mental stimulation can change the lives of both dog owners and their pets. You may realize your dog's full potential and have a more satisfying relationship with your dog by implementing this professional advice and taking cues from real-life success stories.

BONUS

Get your video course here: kindly type in those link on your browser to gain access or you scan the QR codes:

Video one: http://tinyurl.com/yf24ycbw

Video Two: http://tinyurl.com/422h95st